Magnetism and Electricity

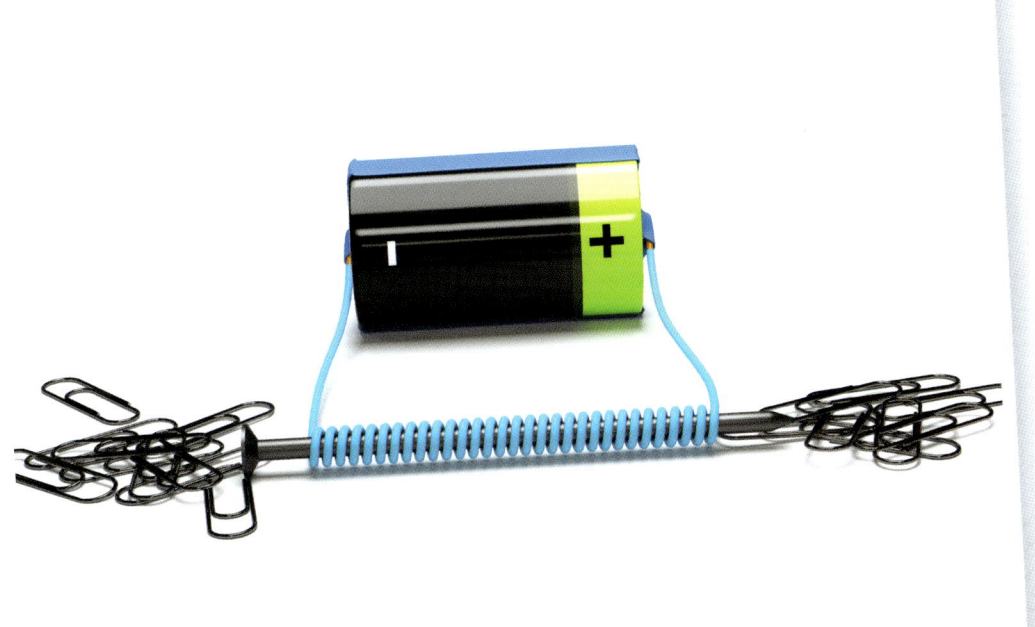

by Emily Sohn and Joseph Brennan

Norwood House Press
For information regarding Norwood House Press, please visit our website at www.norwoodhousepress.com or call 866-565-2900.

Contributors: Edward Rock, Project Content Consultant
Editor: Lauren Dupuis-Perez
Designer: Sara Radka
Fact Checker: Sam Rhodes

Photo Credits in this revised edition include: Getty Images: Cultura RF, 20, iStockphoto, cover, 1, 21, Jeffrey Coolidge, 8, Nastasic, 22, tock, 16 (top), Uppercut RF, 18; Pixabay: Byunilho, background (paper texture), GDJ, background (tech pattern), Spowell, 4; Wikimedia: Daderot, 23

Library of Congress Cataloging-in-Publication Data
Names: Sohn, Emily, author. | Brennan, Joseph K. (Joseph
    Killorin), author. | Sohn, Emily. iScience.
Title: Magnetism and electricity / by Emily Sohn and Joseph Brennan.
Description: [2019 edition]. | Chicago, Illinois : Norwood House Press, [2019] | Series: iScience
    | Audience: Ages 8-10. | Includes bibliographical references and index.
Identifiers: LCCN 2018057852 | ISBN 9781684509614 (hardcover) |
    ISBN 9781684043804 (pbk.) | ISBN 9781684043910 (ebook)
Subjects: LCSH: Electricity—Juvenile literature. | Magnetism—Juvenile literature.
Classification: LCC QC753.7 .S64 2019 | DDC 537—dc23
LC record available at https://lccn.loc.gov/2018057852

Hardcover ISBN: 978-1-68450-961-4
Paperback ISBN: 978-1-68404-380-4

Revised and updated edition ©2020 by Norwood House Press. All rights reserved.
No part of this book may be reproduced without written permission from the publisher.
CIP Information listed above.
341—052021
Manufactured in the United States of America in North Mankato, Minnesota.

Contents

iScience Puzzle	6
Discover Activity	7
What Is a Magnet?	9
What Is an Electric Circuit?	14
Science at Work	20
Connecting to History	22
What Is an Electromagnet?	24
Solve the iScience Puzzle	28
Beyond the Puzzle	29
Glossary	30
Further Reading/Additional Notes	31
Index	32

Note to Caregivers:
In this updated and revised iScience series, each book poses many questions to the reader. Some are open ended and ask what the reader thinks. Discuss these questions with your child and guide him or her in thinking through the possible answers and outcomes. There are also questions posed which have a specific answer. Encourage your child to read through the text to determine the correct answer. Most importantly, throughout the book, encourage answers using critical thinking skills and imagination. In the back of the book you will find answers to these questions, along with additional resources to help support you as you share the book with your child.

Words that are **bolded** are defined in the glossary in the back of the book.

Magnet Fun

Many toys use **magnets** to boost the fun factor. Magnets can push toy trains along. They make some puzzle pieces stick. You may even have magnets on your refrigerator. Why do you think magnets stick to some things but not others? How can magnets make objects move?

In this book, you will learn how magnets work. You will also learn how magnets are related to **electricity**. Finally, you will solve a puzzle. It is your job to fix a toy car.

iScience Puzzle

Fix it

Uh oh! Your little sister broke her toy car. The **switch** broke, and now she can't turn it on or off. Two wires are dangling. There are still batteries in the car, but the car will not go. How can you fix her car?

Here are three ideas:

Idea 1: Use Electricity.
Twist the ends of the wires together. This will make the car go.

Idea 2: Use Magnets.
Attach magnets to the car. Then, use bigger magnets to push and pull the car.

Idea 3: Use Electricity and Magnets.
Wrap the end of each wire around an iron nail. The nails will become magnets that snap together to make the car go.

Which idea would work best? As you read, you'll learn how wires work. You will also learn about magnets. That will help you solve the puzzle.

Discover Activity

Materials
- small magnet
- large magnet
- ruler
- nails of various sizes
- paper clips

Playing with Magnets

Magnets have some unique properties. One is their ability to **attract**. If you hold magnets the right way, they pull some metal objects to them. It can seem like the objects are moving by themselves.

Here's a way to test how strong a magnet is. Push a small nail close to the small magnet. How close does the nail have to be for the magnet to move it? Use a ruler to find out. Record your results in a data table like the one on the next page. How do your results change if you use a larger nail with the same magnet? Try other combinations to complete your data table.

Data Table

Magnet size	Nail size	Distance between nail and magnet when nail starts to move
small	small	
small	LARGE	
LARGE	small	
LARGE	LARGE	

Some refrigerator magnets are stronger than others. How could you find out which magnet is strongest?

Try another activity to compare refrigerator magnets. Make several short and long chains of paper clips. Which magnet will hold a chain with more paper clips? How big is the difference between the number of paper clips each will hold?

What Is a Magnet?

You might use magnets to pull, or attract, other magnets. Or the magnets could push away, or **repel**, each other. These forces that act between magnets or between magnets and metals such as iron are called **magnetism**. How did you see magnetism at work in the Discover Activity?

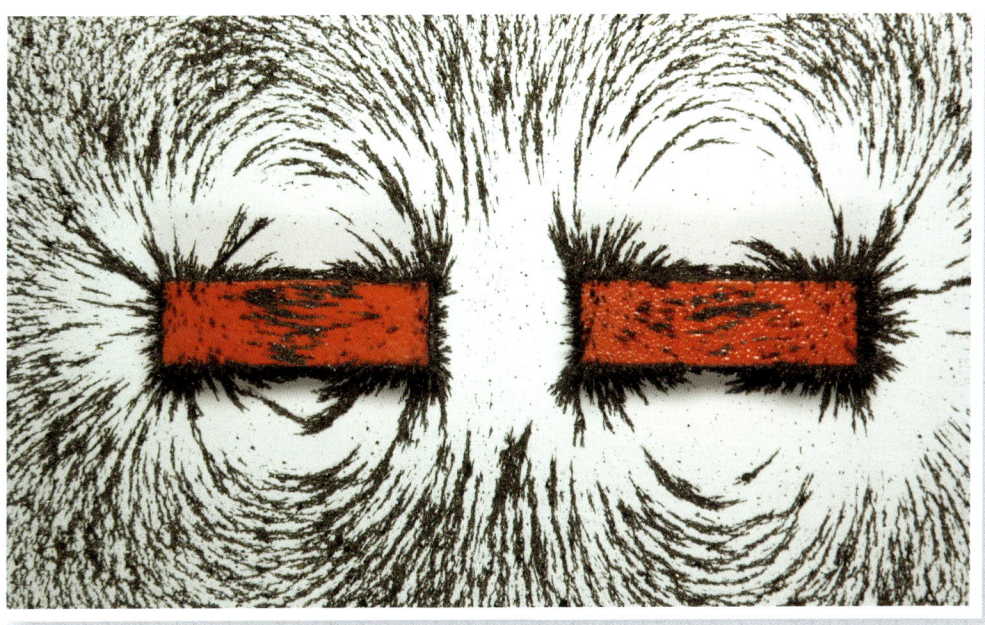

The red bars in this image are magnets. The black specks are little bits of iron. The iron specks line up where the magnetic force is strong. The area of force around a magnet is called the magnetic field.

The magnet looks stuck to this big piece of rock. The rock might be magnetic.

Toy stores aren't the only places to get magnets. You can also find them in the ground in certain places on Earth. Some rocks are magnetic. **Lodestone**, or magnetite, is one example of a magnetic rock.

Magnets can be strong or weak. Think about the refrigerator magnets you tested in the Discover Activity. Then look at idea 2 in the iScience Puzzle. Would a refrigerator magnet be strong enough to pull the toy car?

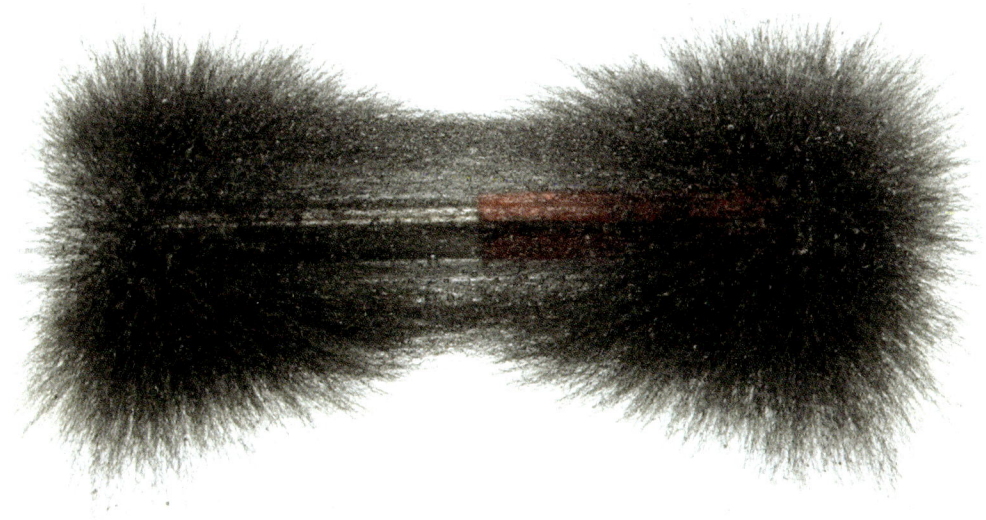

Look at the pattern of iron bits stuck to this red-and-white magnet. The whole magnet has magnetic force. But the forces are strongest at the poles.

Power Source

Every part of a magnet can push or pull. But a bar magnet's force is greatest at its ends. The ends of a magnet are called **poles**. One end is the north pole. The other is the south pole. Both poles have magnetic forces. But the direction of the force at one pole is opposite to the direction of the force at the other pole.

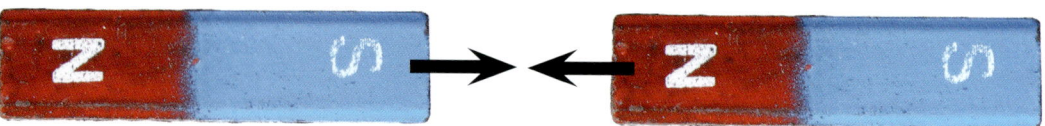

Opposite poles attract each other.

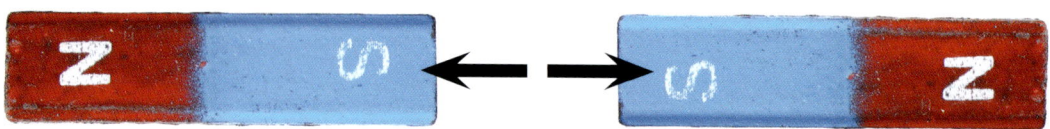

Poles that are the same repel each other.

You may have heard that opposites attract, or pull together. This is true for magnets. The north-pole end of a magnet attracts the south-pole end of other magnets. But poles of the same kind repel, or push away. North poles push away other north poles. South poles push away other south poles.

How do you need to hold the magnets so they attract each other? How would you need to attach magnets on the car to make iScience Puzzle idea 2 work?

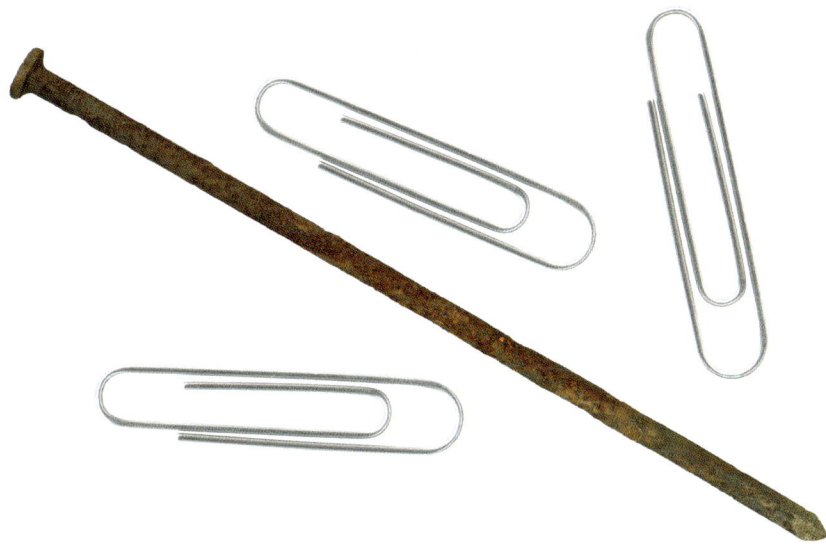

You can turn a nail into a magnet that will pick up a paper clip. It's not magic. It's science!

Magnet Power

Put an iron nail near a paper clip. Chances are, nothing will happen. But a magnet will pick up the paper clip. Now, rub the nail with the magnet. Make sure to rub in just one direction. Do this at least 50 times. Now, see if you can get the nail to pick up the paper clip. Next, test the rubbed nail with the magnet. Does one end of the magnet attract both ends of the nail?

You've just transferred magnetism into a nail. Could you use this technique to help a toy work?

What Is an Electric Circuit?

You've explored magnets as a way to make things move. Now, take a fresh look at the iScience Puzzle. This time, consider electric **circuits**. That's what used to make your sister's toy car work. Electricity moved through it in a complete loop. The switch completed the loop.

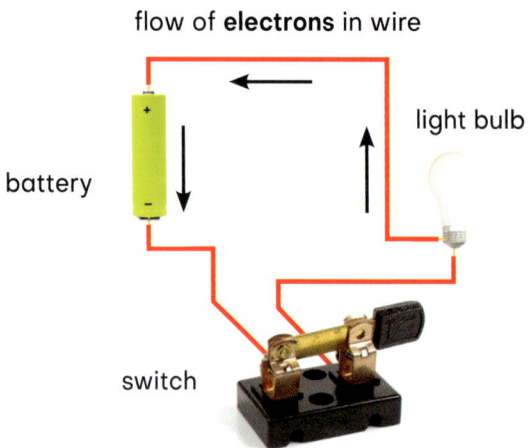

flow of **electrons** in wire

battery

light bulb

switch

When the switch is closed, electrons move around the circuit and light up the bulb.

Electricity usually comes from a flow of small particles called electrons. Electrons have a negative **charge**.

Electrons move through wires only when the wires make a circuit or complete loop. When the toy car's switch broke, the circuit was no longer complete. Electrons stopped moving.

On the Move

A battery has two ends. One end is negative. The other is positive. Suppose a wire connects the negative end of a battery to the positive end. Electrons, which are negative, are repelled by the negative end and are attracted to the positive end. But they flow only when the wire is connected to both ends. This makes a complete loop.

How are forces acting on electrons similar to forces acting on magnets?

A battery's positive end is always marked with a plus sign (+). It has a little bump on it. The negative end is flatter. It is marked with a minus sign (−).

You cannot see the water inside the hose. You cannot see the electric current inside the wire. You can see a current if it jumps a gap between two wires.

In the Flow

Electrons can really move. But they're not always on the go. Certain objects, such as some metal wires, are good **conductors**. Electrons can flow through them easily. This flow is called an **electric current**. It is like water rushing down a river. Other objects do not allow currents to flow through them easily. You've likely never seen wires made from poor conductors, such as cotton or wool.

Think about your sister's toy car. How are the wires in the toy car like a garden hose?

Electrons do something as they flow. They release **energy**. Devices can then use this electrical energy.

Some devices run on batteries, like your sister's toy car. Other devices plug into a wall outlet. The flow of electrons from a wall outlet is produced in a different way than in batteries. But the flow is still an electric current. The current from wall outlets is more dangerous because it carries much more energy than a battery.

Electric currents provide energy to light this lamp. What other devices do you use that run on electricity?

Some devices use rechargeable batteries. These devices run when they are not plugged in. But you have to recharge the batteries by plugging into a wall outlet. What other devices in your home or school use rechargeable batteries?

Which Way Pathways

Some electric currents move like runners on a track. They go along a single path. And they always go the same way. This is called a **series circuit**.

Strings of decorating lights can work this way. Every light on a string can be attached to just one wire. If one bulb breaks, none of the lights in the string will work. The circuit is broken.

If one bulb burned out in a series circuit, all of the other lights would stop working, too.

The toy car in the iScience Puzzle uses a series circuit. The current in the car flows along one path. Without the switch, the circuit is broken. Electrons can no longer flow.

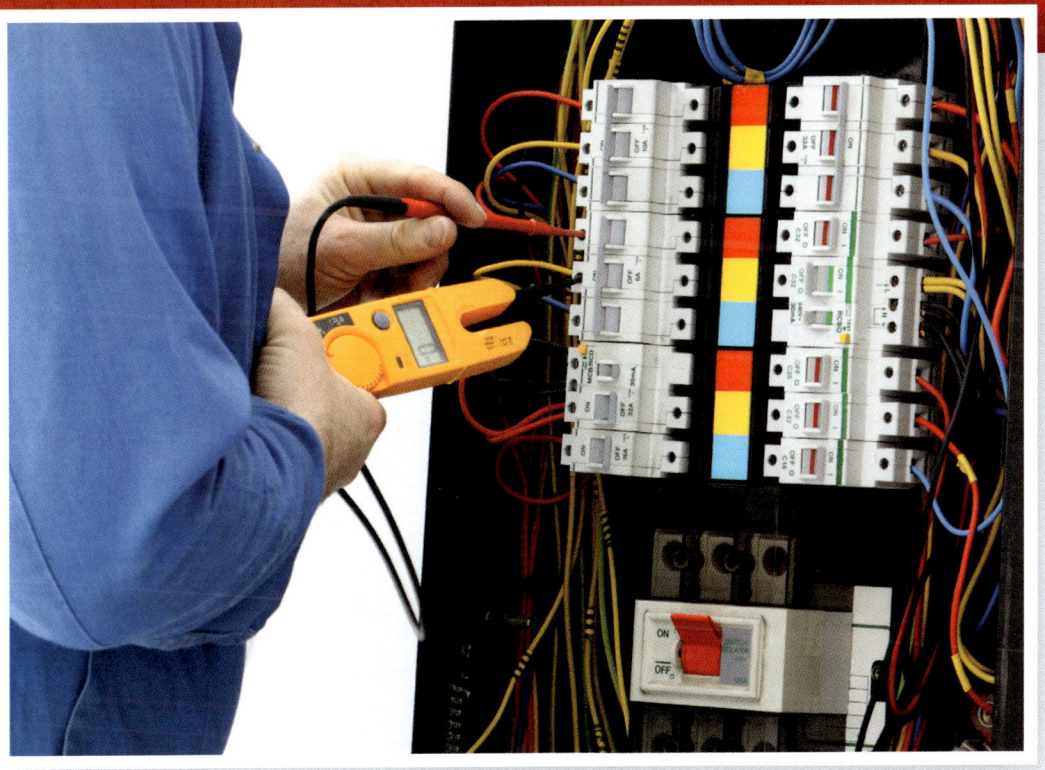

The wires in your home attach inside a box like this one. With many wires, there are many circuits for currents to flow through. CAUTION: In this photo, a protective cover was removed to test the circuits. Only a professional should work inside these boxes.

Not all circuits work this way. Say a light bulb in your home burns out. Chances are, all of the other lights will keep working. Most homes are wired with **parallel circuits**. Electricity can move through several paths. Each plug gets its own loop. Most modern strings of decorative lights are wired in parallel.

Science at Work

Electrical Engineers

Electrical engineers are problem solvers. They start by figuring out what a device should be able to do. Then they figure out how to do it. Electrical engineers design and create. Their projects include cell phones, robots, and computers. They even design wiring systems for entire buildings.

Electrical Engineers can hold many jobs. They can work in offices, out in the field, or in laboratories.

Electrical engineers design ways to make electricity. They might build engines or design turbines. A turbine creates energy from wind or water. The energy is then sent to a generator. The generator powers equipment like lights and computers. Electrical engineers also design how products and systems are controlled. Some systems are automated. Some are controlled by remotes. Some are even controlled by robots!

Electrical engineers design and create everything from video games to jumbotrons in football stadiums. They are constantly improving products and systems to make them better.

Magnetic compass from China

Did You Know?

Long ago, there was no easy way to use electricity. Magnetism was easier to use. Thousands of years ago, people in China made this magnetic **compass**. They used lodestone to make a needle in the shape of a spoon. The handle of the spoon always pointed south. What markings do you think they put on the square plate?

Connecting to History

Faraday: Electric Motor and Electric Generator

Michael Faraday

During the 1800s, many scientists were studying magnetism and electricity. Most people believed electricity moved through wires like water. A chemist named Michael Faraday believed it was a force or vibration. He conducted many experiments to prove his theory. Faraday proved that changes in a magnetic field create an electric force. His research of magnetism and electricity helped him invent two very important devices.

Faraday's first major invention was the electric motor. An electric motor turns magnetism into motion. There are two magnets in a motor. One is a permanent magnet. The other is an electromagnet. Today electric motors are all around us. You can find them in computers, many toys, and even an electric toothbrush.

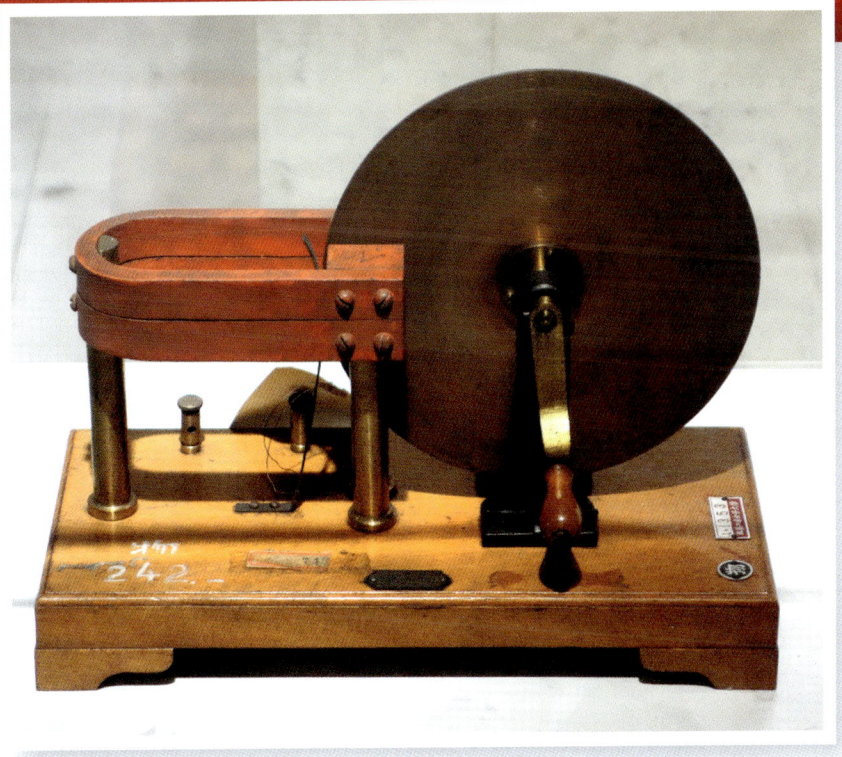

The Faraday disk was the first electric generator. A hand crank spins a copper disk between two magnets and electricity is produced.

Faraday also invented the first generator. A motor uses electricity to make motion. A generator uses movement to create electrical power. A generator can be very small or very large. Generators can be powered by walking, falling water, or by the strong force of the wind. When the turbine begins to rotate, electricity is created. This electricity can be used to power large engines, and even entire factories.

What Is an Electromagnet?

You've looked at magnets and electric currents. You can also make a magnet from electricity. An **electromagnet** shows how the two work together.

To make one, start with metal wire. Wrap it around an iron nail. Attach each end of the wire to one end of a battery. This forms a circuit. Electrons flow through the wire. The nail acts like a magnet.

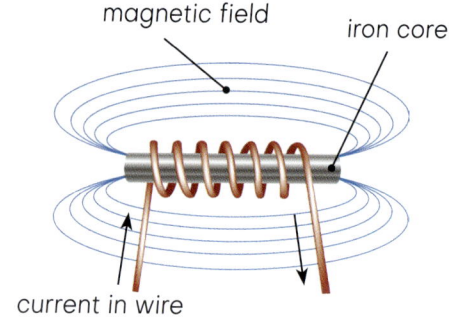

An electromagnet is a wire wrapped around a core made of iron or some other metal that can be magnetized. When current flows through the wire, it causes the core to become a magnet.

Electromagnets can be useful. You can turn them on or off. You can make their current strong or weak. Or you can change the direction of the current. This changes how the magnet acts.

Put paper clips near the wrapped nail with current flowing. What happens? What happens to the paper clips if the circuit is broken? Save your electromagnet. You'll use it again later. (Make sure to disconnect the battery.)

Look at the iScience Puzzle choices. Would it be better to use magnets or electricity to make the car go? Why? How might an electromagnet help you solve the problem?

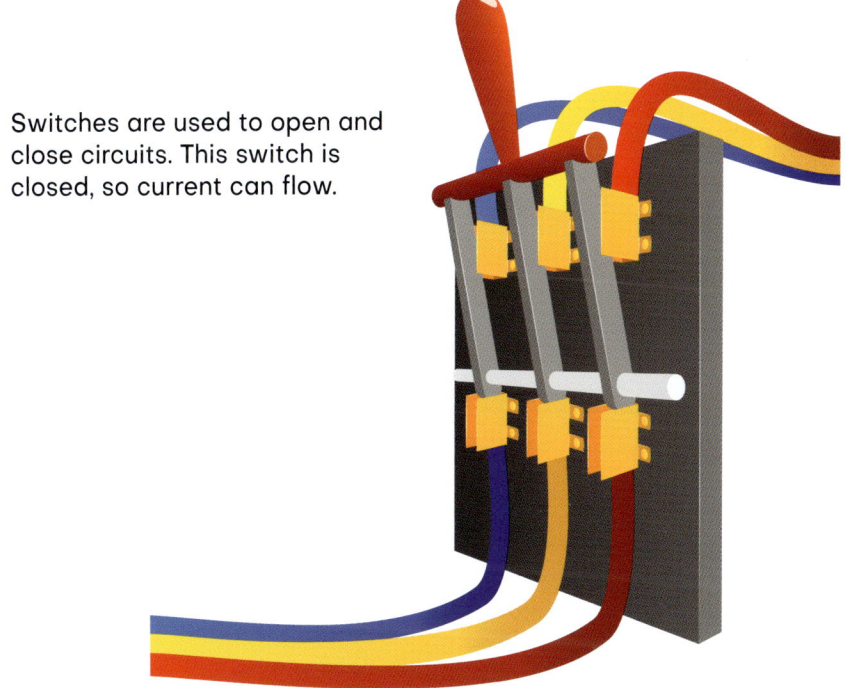

Switches are used to open and close circuits. This switch is closed, so current can flow.

Here's one thing to think about. How will you turn the car on and off? The car had a switch. Moving the switch one way connected the wires. Moving the switch the other way pulled them apart. Could you add a switch to any of the puzzle choices?

Wires can act like magnets, too, even when there is no iron core. That happens when electrons flow through them. Electric currents can flow one way or the other.

Look at idea 3 of the iScience Puzzle. A wire loops around a nail. If current flows through the wire, those loops make a magnetic force. Turn back to your electromagnet. Try adding more coils to the wire. How many paper clips can the nail pick up with 2, 6, or 10 coils wrapped around the nail? Does this give you any ideas about the iScience Puzzle?

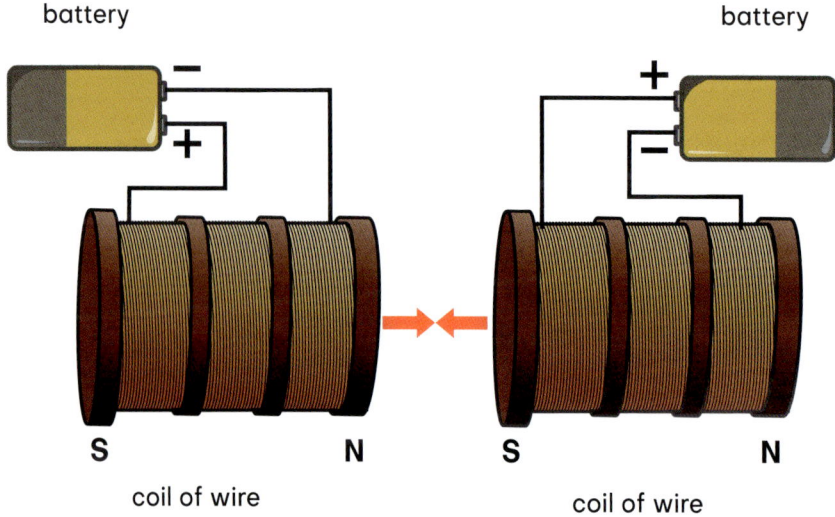

When the current flows in each coil, the coils act like magnets. The opposite poles attract, so the coils move toward each other.

Solenoids are a special type of electromagnet. They have wire coils that are wound tightly. Just like in other electromagnets, current flows through a coil of wire. The coil is wrapped around an iron core. This turns them into magnets.

The difference with a solenoid is that the iron core moves in and out of the coil of wire. This allows them to push or pull other parts in a machine.

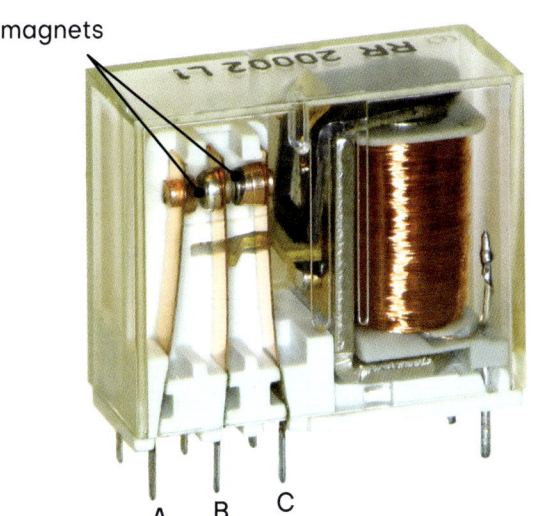

This solenoid can control the flow of current through A, B, and C.

Solenoids also can open and close switches. They show up in doorbells, printers, and other devices. Do you think there are any solenoids in your home or school?

27

Solve the iScience Puzzle

You have learned a lot about magnets and electric currents. Now, which idea do you think would work best for fixing your sister's toy car?

Idea 1: Use Electricity.

This would make the toy car work. Adding a switch would allow you to turn the car on and off, too.

Idea 2: Use Magnets.

This might work. But you'd need a huge magnet to pull and push the car. Also, you would have to run along with it to make it move.

Idea 3: Use Electricity and Magnets.

A wire wrapped around an iron nail will act as a magnet only if current flows. Maybe you could use this idea to make a switch.

Idea 1 might be the best way to fix the car. You might have other ideas now, too. Could you add new features that would make the car even better? Way to go! You've made your sister very happy.

Beyond the Puzzle

Workers at scrap yards often use electromagnetic cranes to move big chunks of metal material.

You have learned about electricity. You have also learned about magnets. You used what you learned to fix a toy car. Now, explore what else electromagnets can do. How can they move the critical parts of motors? How do they help doctors? Some electromagnets are very big. Machines use them to pull iron out of the ground.

Why do you think people make electromagnets? After all, they could just use chunks of lodestone.

Now, see if you can come up with a science fair project. You can use magnets, electric currents, or both. See how creative you can be!

Glossary

attract: to pull toward.

charge: a property of matter that determines how the matter is affected by electric force.

circuits: loops of wire that allow electricity to pass through.

compass: a device with a floating magnetic needle that points to the magnetic north of Earth.

conductors: materials through which current flows easily.

electric current: a flow of charged particles, usually electrons.

electricity: the force that acts on charged particles.

electromagnet: a magnet created by the flow of electricity around a metal core.

electrons: tiny particles that have a negative charge.

energy: the ability to cause changes.

lodestone: a kind of magnetic rock found in Earth.

magnetism: the force that acts on magnets and some metals.

magnets: pieces of metal that can attract or repel some other metals.

parallel circuits: electric circuits with more than one path for current.

poles: the opposite ends of a magnet.

repel: to push away.

series circuit: electric circuit with only one path for current.

solenoids: coils of wire with an iron core that moves in and out. They are often used as a switch in a circuit.

switch: a device that turns something on and off.

Further Reading

Urquhart, Kristina. 2019. *Making Music with Magnets.* Huntington Beach, Calif.: Teacher Created Materials.

Connors, Kathleen. 2018. *Magnetism.* A Look at Physical Science. New York: Gareth Stevens Publishing.

Kenney, Karen Latchana. 2018. *Magnetism Investigations.* Key Questions in Physical Science. Minneapolis, Minn.: Lerner Publications.

Forest, Christopher. 2018. *Focus on Magnetism.* Hands-On STEM. Lake Elmo, Minn.: Focus Readers.

Additional Notes

The page references below provide answers to questions asked throughout the book. Questions whose answers will vary are not addressed.

Page 8: The strongest magnet will pick up the most paper clips.

Page 10: The refrigerator magnet would be too weak.

Page 12: To pull the car with a magnet, face its north pole toward the south pole of a magnet attached to the front of the car.

Page 13: No. One end is repelled by the magnet.

Page 15: Electrons can be attracted or repelled. Magnets can be attracted or repelled.

Page 16: Current flows through the wires in the toy car just as water flows through a hose.

Page 21: Did You Know? question: They probably marked the plate with north, south, east, and west.

Page 24: When current is flowing, the iron core will attract paper clips. When current stops flowing, the iron core will drop the paper clips.

Page 25: A switch could be added to idea 1 and idea 3.

Index

attract, 7, 9, 12, 13, 15, 26

circuits, 14, 18, 19, 24, 25

compass, 21

conductors, 16

electric current, 16, 17, 18, 24, 25, 26, 27, 28, 29

electricity, 5, 6, 14, 17, 19, 20, 21, 22, 23, 24, 25, 28, 29

electrons, 14, 15, 16, 17, 18, 24, 26

energy, 17, 20

Faraday, Michael, 22, 23

lodestone, 10, 21, 29

magnetic field, 9, 22, 24

magnetism, 9, 13, 21, 22

magnetite, 10

magnets, 5, 6, 7, 8, 9, 10, 11, 12, 13, 14, 15, 21, 22, 23, 24, 25, 26, 27, 28, 29

parallel circuits, 19

poles, 11, 12, 26

repel, 9, 12, 15

series circuit, 18

solenoids, 27

switch, 6, 14, 18, 25, 27, 28